The Giza Clock and Power Plant

Disclaimer

46 St. Books
Published by 46 St. Books
46 St. Books USA Inc.,
Philadelphia, PA 19144, USA

Copyright © William C. Henry Sr., 2014

All rights reserved under International and Pan-American Copyright Conventions.

ISBN 13# –978-0-9916520-4-4   ISBN 10# - 09916520-4-4

Library of Congress Cataloging-in-Publication Data

LCCN#

Henry Sr., William.
The Giza Clock and Power Plant

PRINTED IN THE UNITED STATES OF AMERICA

PUBLISHER'S NOTE
This book is a work of non-fictional based on archival scriptures, texts, and documents. All images used are either owned, work for hire, acquired works, use with permission, and from the Library of Congress.

BOOKS ARE AVAILABLE AT QUANTITY DISCOUNTS WHEN USED TO PROMOTE PRODUCTS OR SERVICES. FOR INFORMATION PLEASE WRITE 46 ST. BOOKS, 55 WEST PENN ST, PHILADELPHIA, PA 19144 OR VISIT OUR WEBSITE AT WWW.46STBOOKS.COM

Dedication

I would like to say thanks and dedicate this book to all those that I can think of that may have influenced my thoughts, expansion, understanding, and learning as a perpetual student. I give thanks to minds such as; Edward Leedskalnin, Nikolai Tesla, Jean-Pierre Houdin, Albert Einstein, Edgar Allen Poe, Phoebe Snow, Pablo Picasso, Sam Cooke, Khalil Ghibran, the Dali Llama, Deepak Chopra, Professor Bernice Henry, Marcus Joe, Khemet (Land of the Black), Paulo Coelho, Luciano Pavarotti, Enrico Caruso, Stanley Kubrick, Freidrich Nietzsche, Neil deGrasse Tyson, Michio Keku, Philadelphia Public Library and School System, Chestnut Hill College, the Easy and Mean streets of Philadelphia, and so many more things, peoples, and places to name at this time. I thank Creation for all of those beautiful souls that I have encountered, and for all of the angry souls as well. May we all begin to rethink, relearn, and re-understand, to avoid any more regrets.

IWALYT

Table of Contents

Introduction - page 9

Removing cultural bias - page 11

The Giza Solar Clock - page 13

The Sun's path pre 2344 BCE - page 21

360 & 360 Day differences and holidays - page 23

5.5°s of damage - page 25

Events since recorded memory - page 27

Who wants to be a pharaoh? - page 33

Egyptian Pharaoh's list - page 37

Star representations and alignments - page 53

Pyramid facts - page 55

The Sphinx is a dog! - page 59

Giza power plant live - page 61

World culture of squaring the heavens - page 65

Global pyramid culture - page 73

Conclusions - page 77

Introduction

How does one begin to propose an understanding so dissimilar to the prevailing opinions of the times and be heard? The world of academia tends to shun the voice of outsiders, thereby silencing their truth to a larger audience. Our faith based systems will circle the wagons in defense of their paradigms even in the face of an incontrovertible truth. The media outlets that have been created to keep the public informed, are often witting or unwitting tools of disinformation. This task of epic difficulty is just what I and others face in today's world. Symbolism reigns over substance leaving reason no rule.

I will show the reader a clearer picture of precisely how the Great Pyramid at Giza was used as a solar calendar. I will also give the reader a picture of how the primary purpose of the complex was electricity generation. This commonly shared theory has gained ground in recent years, as this was never a funerary complex, and did not arise from the decision to expand a mastaba. The prevailing biases at the time the complex was rediscovered up to now, have led to a false portrayal of the buildings significance. The obfuscation of truth and knowledge becomes all the more reprehensible when it limits one's mind and soul.

Removing Cultural Bias

When the blossoming discipline of what has become known as "Egyptology" began, the European archeologists infused their ethnic and cultural biases into their marvelous works and discoveries of lost cultures and knowledge. The early French and English archeologists among other ethnic groups, constantly portrayed the Egyptians as lacking a math system and being preoccupied with death and their own mortality. The pyramids, statues, and living quarters, were carefully laid out for solar and celestial alignments to aid in power generation.

The pyramids at Giza NEVER held the body of a pharaoh or anyone else! This mistake based on ethnic biases at the time, has caused the true use of the pyramids to be smudged of accuracy. The builders of these magnificent worldwide structures were of a more advanced intelligence than we are at this point in time. The debate has raged in recent years as to the purpose of the pyramids? Many researchers in various fields have come to the conclusion based on all available material, that the pyramids were in fact designed for power generation and distribution of some form. The Giza complex was a power factory and so much more. Coupled with the outstanding research of others, and my own diligent work, I have come to see the complex in such a greater and more enlightened way.

The pyramid complex at Giza was not conceived of, designed by, or built by slaves! The Torah and Biblical accounts relate the Hebrew people as helping to build the pyramid complex at Giza. The pyramid complex was built 600 to 800 years prior to the entrance of these people into Egyptian society. The Giza complex was built historically around 2560 BCE, and my personal alignment dating is 2740 BCE. The Hebrew people do not enter the land of Egypt according to their own eventual written accounts until around 1900 BCE. The Hebrew people do not have a written language at this point in their history, and this writing system will not be solidified until around 600 BCE.

The Hebrew people did not help to build the pyramid complex, but were possibly used as laborers to help dig the complex out from under the newly formed deserts. I will layout later in greater detail the 2344 BCE date for the Passover event of the Great Global Flood. Scientists have recently dated the first deserts on the planet as forming around 2300 BCE. This is but one item that supports my datings and overall view of events. The ongoing Earth changes that occur after the event causes drought, famine, war, and upheavals around the globe. The Flooding of the Black Sea region forces peoples into other lands seeking regions more suitable for their resettlement. The Hebrew people appeared in Egypt around 1900 BCE, and resided there until the disasters that would eventually be related Biblically as the Exodus Disasters around 1500 BCE to 1487 BCE.

The Giza Solar Clock

The Great Pyramid was designed as a power plant and a solar/cosmic timepiece. Once we come to terms that the pyramids were designed for these purposes, and oriented to the east toward the rising sun understanding becomes easier. This eastward orientation is encouraged in the Oriental practice of Feng Shui, in the positioning of the body for prayer in the Islamic faith, and with the Jewish and Christian faithful at the Western Wall in Jerusalem, that is kissed while facing east! This author is not here to criticize a faith, but to show the true understanding in a practice or tenet.

The Egyptian system as is our own, was based on the 360 degrees in a circle. The use of the 360 count system is the manner that the celestial heavens were mapped, and this was also the format used in the calendar and math systems. The pyramids were aligned as early as 2740 BCE (2740 BCE - 2344 BCE = 396 years ÷ 72yrs = 5.5 celestial portions!) and as late as 2560 BCE for the building. The super-scribed circumference of the base of the Great Pyramid equals the speed of light (186,000mi/sec) in a vacuum. The GPS coordinates for the interior of the King's Chamber is 29 degrees 58 minutes 49 seconds north latitude by 31 degrees East 09 minutes 0 seconds of arc or 29.98027° equal to 299.8M or 299,792,458m/s or the speed of light squared.

There are many more tie ins to the later system of Pythagorean math that formed through usurping prior knowledge, and with ethnic replacement came to supplant the teachings of a much older and more knowledgeable civilization. The Pythagorean equations known as Pi and Phi are incorporated into all of the building aspects of the Giza complex, as well as the Golden Ratio of 1 to 1:618! These were not coincidences or mistakes as alluded to by many researchers and authors. Disciplines such as archeology and mathematics are based on a set of empirical standards that should be free of bias based on ethnicity or religion.

When we understand that the Pre-Flood (2344 BCE) calendar was setup for 360 days versus our 365 day calendar, we are able to align more events and true understanding. A 360 day calendar would give us 12 months of 30 equal days each. The axis of the Earth before the Global Flood was 18-degrees, as opposed to our current 24 hour day. This partially explains why the pre-Flood people lived extremely long lives in one sense. The chart below shows the relationship of the 18 hour day and 360 day calendar system to give a clearer picture.

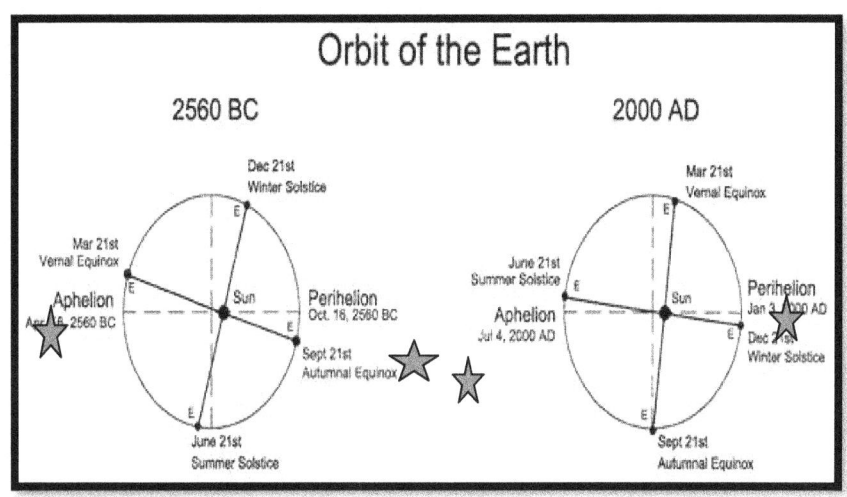

The aphelion and perihelion points of the Earth changed after the devastating Planetary Passover of April 16, 2344. The aphelion occurred on 4/16 at the 106° mark and shows a 79 day difference with the 2012 aphelion of 7/4 at the 185° point. The former perihelion occurred on 10/16 at the 286° mark and shows a 77 day change with the current perihelion point of 1/3 at the 3° point. The mean difference between the 77 days and 79 days would be 78 days. The seasons changed overnight as the Sun was at aphelion allowing for easy passage for the Planetary Passover of Venus. Immanuel Velikovsky most notably put this theory forward, and my material supports a view of periodic cosmic destruction.

18 HOUR DAY VS 24 HOUR DAY BY DESIGN?

18 Hr Day / 360 Day	Seconds	Minutes/Hours	24 Hours
Minute	1 secs	60 secs = 1	60 secs =

		min	1 min
Hour	60 mins hr X 60 secs = 3,600 secs hr	60 mins hr X 360° = 21,600 secs (2160yrs = Zodiac Age) * 2160 X 12 Ages = 25,920 yrs = A complete Zodiac Cycle!	60 mins X 60 secs = 3,600 secs hr
18 Hr. Day	18 hr day X 3,600 secs hr = 64,800 secs day	18 hrs X 60 mins = 1080 minute day * 1080 resolution! Curious? 86,400 - 64,800 21,600 216° Solar point 2160 Zodiac Age	60 mins X 24 hrs = 1440 min day 86,400 secs day Speed of Light = 186,400' sec *3,600 X 12 = 43,200?
Week	64,800 secs day X 7 days = 453,600 secs week 1 year = 6,480 hrs	18 hr day X 7 days = 126 hours week	86,400 secs day X 7 days = 604,800 secs wk 24 hr day X 7 days =

			168 hrs wk
30 Day Month	453,600 secs Week X 4 = 1,944,000 secs mo	18 hr day X 30 days = 540 hours mo	604,800 secs wk X 4 = 2,419,200 secs mo 24hr day X 30 = 672 hours *5.5 days added to calendar
Year	1,944,000 secs mo X 12 = 23,328,000 secs yr	540 hr mo X 12 mo = 6,480 hr yr * 6,480 ÷ 3 = 2,160 = A Zodiac Sign *6,480 X 4 = 25,920 = A complete Zodiac Cycle! 540 X 2mo = **1080**hrs	2,419,200 secs mo X 29,030,400 secs yr 672 hr mo X 12 mo = 8,064 hrs yr
Pharaoh's 50 Year Jubilee	23,328,000 sec yr = 1,166,400,000 secs 50 yrs	324,000 hr yr ÷ 50 yrs = 6480 hr yr * 400 Yr. Jubilee ÷ 8	

		sides = 50 yrs per side	
Tropical/Solar Year = 400 ∞ Great pyramid has 8 sides = 50 years per side	400 yrs X 6,480 hrs yr = 2,592,000 hr Jubilee * 2,592,000 hr ÷ 360° = 7,200 ÷ 18hrs = 400!!!	Note: the 2,592,000 equals the mean astronomical distance of .0026au of the Earth and Moon	394.26 Trop Yr ÷ 8 = 49.2825

18 Hours vs 24 Hours (+) (-) or (=)

Minutes: 18 & 24

(+) Hours 18 - The 3600 seconds per hour relates to the 360 degrees in a circle (360 X 60 = 21,600) and 360 calendar days

(+) Minute per day 18 - There is 1080 minutes per day equal to 1080 resolution

(+) Minutes per day 18 - The 24hr day seconds 86,400 - 64,800 18hr day seconds = 21,600! This is the difference in amount to the system that was adjusted from 18hrs. to 24hr. days

(+) Week 18 - A week of 18hr day seconds X 12 months = 6,480hrs. which equals 1/4 of a Great Zodiac Cycle of 25,920 years

(+) 30 Day Month 18 - The numbers are not outstanding, but the 24 hour day needs intercalary days (5.5) added during the year

(+) Year 18 - A year of hours (6,480) equals 1/4 of a Great Zodiac Cycle of 25,920 years (6480 X 4 = 25,920 ÷12 = 2160yrs = A Zodiac sign!

(+) Pharaoh's 50 Year Jubilee 18 - Fifty years of 324,000hrs ÷ 50yrs = 6,480hrs per yr.   *400yrs ÷ 8 sides = 50 yrs per side

(+) Tropical Year 18 -

144,000 days ÷ 360 = 400yrs

6,480hr yr X 400 = 2,592,000hr = the mean astronomical distance between the Earth and Moon of .0026au

*144,000 days ÷ 365 = 394.52054 Tropical Years

The Egyptian system of timekeeping placed no open values for decimals and fractions of numbers, and this would speak to all things in their system being based upon the 360°s in a circle!   18 = 8   24 = 0

An aerial view of the Great Pyramid shows the concave and 8 sided feature to the pyramid. The pyramid lines are at the 324° East, 72° South, 144° North, and 252° North respectively.

SOLAR SEASONAL SIDES

East Side: Spring to Summer

South Side: Summer to Fall

West Side: Fall to Winter

North Side: Winter to Spring

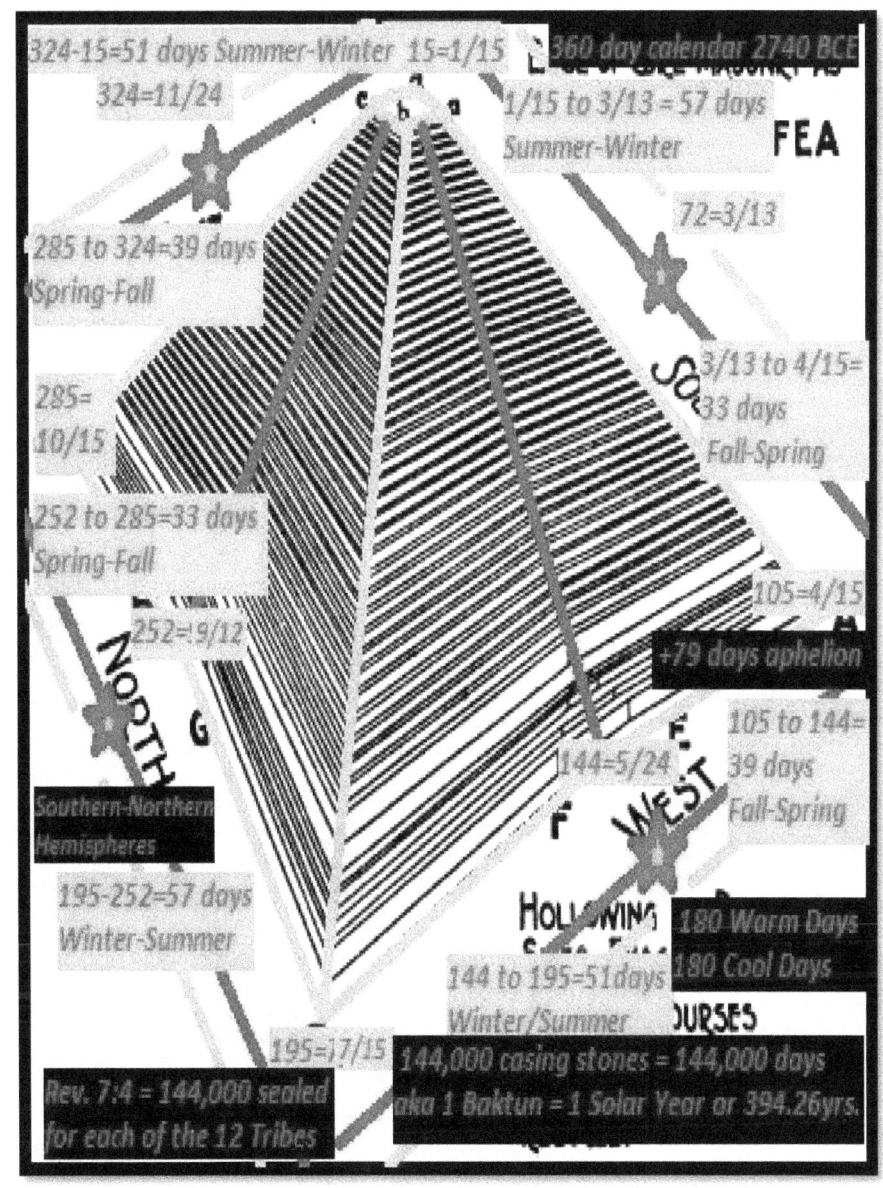

∞ The exterior of the Great Pyramid of Giza was originally lined with 144,000 massive casing stones from 5 to 30 tons each. This system was used to count periods of 144,000

days known as a Tropical Year, and as a Mayan Baktun. The Mayan system and the modern Tropical/Solar Year both equal 394.26 years (394.52yrs?) based on a 365 day year. This loss of 5.74 days almost mirrors the loss of 5.48yrs or the addition of 5.5 calendar days after 2344 BCE.

Pyramid: 144,000 days ÷ 360 days = 400 years ÷ 8 sides = 50 years each Pharaoh's Jubilee.

Baktun/Tropical: 144,000 days ÷ 365 days = 394.26 years ÷ 8 sides = 49.2825 years each side

The Mayan Calendar Agrees with Egypt

Days	Long Count Unit	Long Count Period	360/365 Day	Diff
1	1 Kin		1	
20	1 Winal	20 Kin	20	
360	1 Tun	18 Winal	360/365	+5
7,200	1 Katun	20 Tun	20yrs/19.72yrs	+.28
144,000	1 Baktun	20 Katun	400yrs/394.52yrs	*5.48yrs
2,880,000	1 Piktun	20 Baktun	8000yrs/7890.42yrs	109.58rs
57,600,000	1 Kalabtun	20 Piktun	160,000yrs/157808.21yrs	2,191.79 yrs
1,152,000,000	1 Kinchitun	20 Kalabtun	3,200,000yrs/3,156,164.38yrs	4,3835.7 yrs
23,040,000,000	1 Alautun	20 Kinchitun	64,000,000yrs/63,123,287.67yrs	876,712.33yrs

The original Mayan calendar shows a synthesis with the Egyptian 360-day calendar. Both systems had no use for fractions for the sum of an equation!

*A 5.48 year value relates to the additional 5.5 days added

The Sun's Path Pre 2344 BCE

Aphelion	Date	Degree	Perihelion	Date	Degree
2740 BCE	4/16	106°	2740 BCE	10/16	286°
2000 BCE	7/4	185°	2000 CE	1/3	3°
	+79	79°		+77	77°

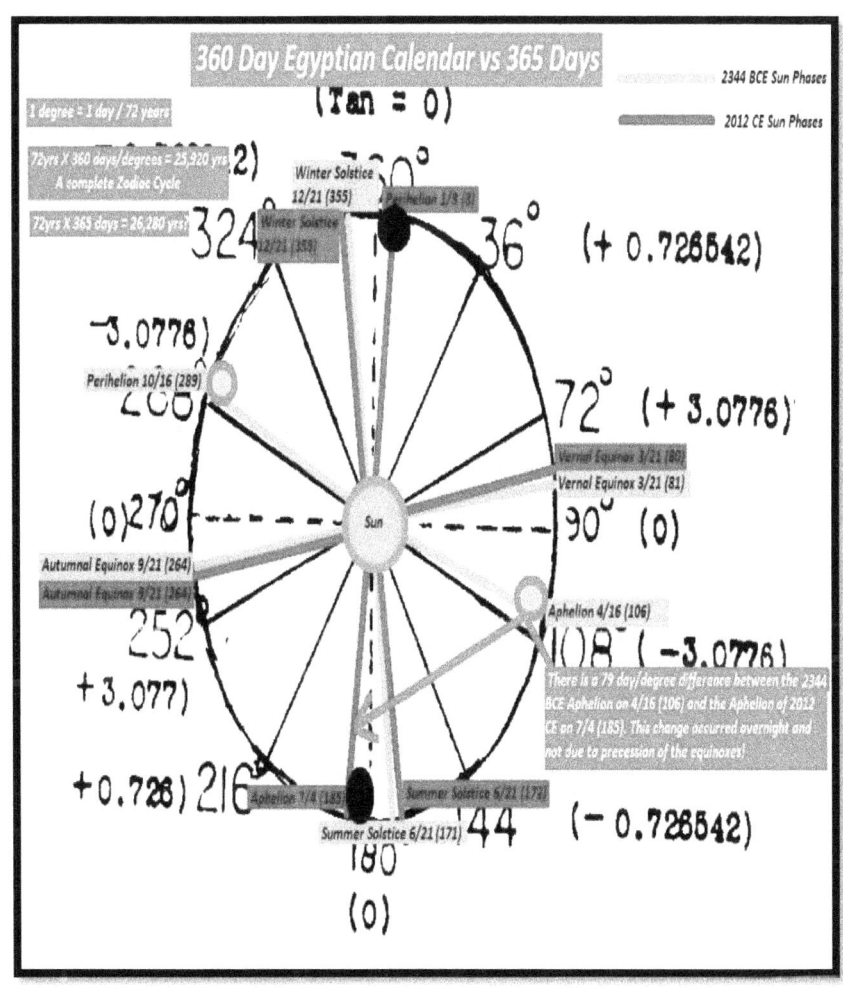

Degree	Day	Solar Point
15°	1/15	
36°	2/6	
72°	3/13	
81°	3/21	Vernal Equinox
105°	4/15	Passover
106°	4/16	Aphelion
144°	5/24	
171°	6/21	Summer Solstice
195°	7/15	
252°	9/12	
261°	9/21	Autumnal Equinox
285°	10/15	
286°	10/16	Perihelion
324°	11/24	
351°	12/21	Winter Solstice

* The Sun's true path prior to 4/15/2344 BCE: 72° and 252°

Degree	Direction	Date	Season	Days	Curious
15°	SE	1/15	Summer	51	15° mag diff
72°	S	3/13	Summer	57	Solar point 72° - 90° = 18° axis
105°	SW	4/15 (A)	Fall	33	4/16 Passover Point
144°	W	5/24	Fall	39	
195°	NW	7/14	Winter	51	
252°	N	9/9	Winter	57	Solar point
285°	NE	10/15 (P)	Spring	33	10/16 Perihelion
324°	E	11/24	Spring	39	

360 & 365 Day Differences & Holidays

2740 BCE degree	2012 CE degree	2740 BCE date	2012 CE date	Degree & Day Diff	Holiday
3°	3°	1/3	1/3 (P)	3° / 0	New Years Day, Solemnity of Mary
33°	32°	2/3	2/2	1° / -1	Imbolc
44°	45°	2/14	2/14	1° / -1	St. Valentine's Day
81°	80°	3/21 (VE)	3/21 (VE)	1° / +1	Vernal Equinox Day, March Equinox Day, Ostara
91°	91°	4/1	4/1	0° / 0	Feast of Kings, April Fools Day,
100°	100°	4/10	4/10	0° / 0	Ascension of Christ
106°	105°	4/16 (A)	4/15 (A)	1° / +1	Passover, Tax Day, Bengali New Year, Easter
121°	121°	5/1	5/1	0/0	May Day, Beltane, Spring Labor Day
171°	172°	6/21 (SS)	6/21 (SS)	1° / -1	June Solstice Day, Litha

185°	185°	7/5	7/4 (A)	0° / +1	Independence Day
261°	264°	9/21 (AE)	9/21 (AE)	3° / -3	Mid-Autumn Festival, Feast of Mabon, September Equinox Day, Autumnal Equinox Day
286°	289°	10/16 (P)	10/13	3° / +3	Hajj, Eid-al-adha
351°	355°	12/21 (WS)	12/21 (WS)	4° / -4	Christ Death, Tammuz Death, Yule Day, December Solstice Day, Christmas - Christ Reborn, Tammuz Reborn
360°	365°	12/30	12/31	+5° / +5 days	365 day calendar +5.5 days

5.5°'s of Damage!

5 Degrees of Separation
Pre 2344 BCE - Earth's axis 18° & Post 2344 BCE - Earth's axis 23.5° = +5.5° and days
King's Chamber vault crack - 5° from NE direction
Earth - Moon sits at 5° inclination was 0° prior to 2344 BCE which supports the Earth - Moon capture theory.
Pyramid alignment 2740 BCE - 2344 BCE = 396yrs / 72yrs = 5.5
2740 BCE Tropical Year = 400 years @ 2344 BCE Tropical Year = 394.52   400 - 394.52 = -5.48 years

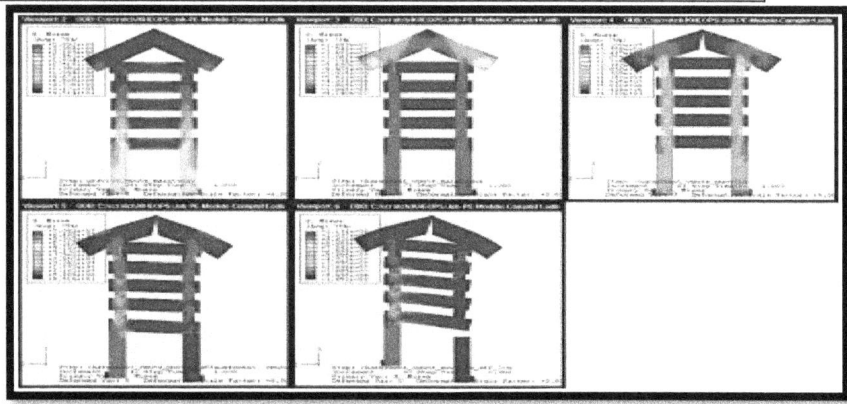

Thermal imaging has shown the vault in the King's Chamber to be cracked at 5 degrees leaning from North to South, as the Planetary Passover took a Northwest approach adding 5.5 days to the current calendars.

Earth Axis 2740 BCE	Earth Axis 2012 CE	Magnetic Difference
0° to 18° = 18°	0° to 23.5°	2740 BCE - 15
18	23.5	2012 - 9
288° ÷ 24 = 12		

	months		
	288° ÷ 12 = 24 hours		

Events Since Recorded Memory

Year	Event	Significance
4000 BCE	Green Earth Sumeria	No deserts on the planet with a possible Pangean continent
3200 BCE	1st known Egyptian writing	
3114 BCE	Start of Mayan time	13 Baktun cycles of 394.26yrs ended in 2012
3102 BCE	Start of Hindu Kali Yuga Age	The special 10,000 year Kali Yuga Age ends in 6898 CE
2740 BCE	Pyramids of Giza aligned to the stars	

2600 BCE	Stonehenge aligned	
2460 BCE	Traditional pyramid building	
2450 BCE	1st Egyptian obelisk built	*Symbolizes the privy member in the sky*
2350 BCE	Pharaoh's Jubilee obelisk built	*Symbolizes the privy member in the sky - Vatican, Rome*

	Massive New Madrid earthquake	An earthquake in North America
2344 BCE	Global Flood Passover event	Worldwide flood of record and religious texts
2300 BCE	1st Deserts appear globally	44yrs after the planetary passover that decimates the Earth
2184 BCE	80% Egyptian cities abandoned Nile River ceases to flow for 100 years Sodom & Gomorrah disasters Akkadia	The land is decimated yet again during the time of the Biblical figure of Abram

	collapses	
1917 BCE	*Hebrews enter the land of Egypt*	*430yrs before the end of their eventual Exodus*
1500 BCE	*Dwarka India sinks into the sea Santorini sinks into the sea*	*Holy city of Lord Krishna Thera in Greece erupts?*
1487 BCE	*Hebrew Exodus from Egypt*	*Egyptian & Biblical planetary pass-over disasters continue*
1447 BCE	*Hebrew settlement in Canaan*	

| 1400 BCE | Libyan Desert Glass used by Pharaohs | King Tut pectoral ornament Cairo Museum @ 1330 BCE  |

Mankind reawakens around 4000 BCE after another series of catastrophes, and this period is remembered as Adapa and Titi in Sumerian lore, and as the period of Adam and Eve Biblically. The Egyptian, Hindu, and Mayan civilizations start their calendar systems around 3100 BCE each. The pyramids are aligned in 2740 CE as construction begins at Giza. The Global Great Flood of 2344 BCE brings a fond remembrance for the Age of Taurus that ends in 2309 BCE, and a lack of recognition for the new Age of Aries that entered in 2308 BCE. Scientists have recently concluded that the first deserts did not appear on Earth until 2300 BCE. This dovetails nicely into the timelines that my research had previously established. I have consistently questioned why we have deserts? And how does sand exactly form? A Planetary Passover by another charged body would bring the resulting land and climate changes. The period of 2184 BCE brings continued disasters as the Nile ceases to flow for 100 years, 80% of Egyptian cities are abandoned, and the Biblically related disasters of Sodom, Gomorrah, Bela, Admah, and Zoar, take place during this time. There comes a period of recovery, until another series takes place in 1500 BCE to 1447 BCE, as Thera and Dwarka sink into the sea, and the Exodus disasters unfold. The material that is discovered by travelers in the newly formed deserts is linked to the God's in the sky that formed it. These pieces of plasma blasted sand became a favorite of Pharaohs to wear as ceremonial jewelry. We call this jewel Libyan Desert Glass, and this type of glass can only be recreated through a nuclear explosion.

Who wants to be Pharaoh?

Egyptian Dynasty	Date	Noted Pharaoh
Early Dynastic	3100 BCE - 2686 BCE	3150 BCE - Scorpion King 3050 BCE - Hor Aha *2740 BCE - Khasekhemwy
Old Kingdom	2686 BCE - 2181 BCE	2600 BCE - Sneferu 2450 BCE - Neferefre 2350 BCE - Unas +2345 BCE - Unas dies 2345 BCE - Teti *2344 BCE - Global Great Flood* 2300 BCE - Meryre Pepi ¤2184 BCE - Merenre Nemtyemsap II ¤2184 - 2181 BCE Neitiqerty Siptah
1st Intermediate	2181 BCE - 2055 BCE	2181 BCE - 2169 BCE 14 Pharaohs - 8 mo. avg.
Middle Kingdom	2055 BCE - 1650 BCE	1917 BCE - Nubkaure Amenemhat II
2nd Intermediate	1650 BCE - 1550 BCE	
New Kingdom	1550 BCE - 1069 BCE	≠1487- 1447 Exodus Disasters Hatshepsut & Tuthmosis III

The cosmic disasters run parallel with the length of reign for the Pharaohs. This is displayed in the period length for each Egyptian Dynasty and rule of those Pharaohs.

The Early Dynastic period of 3100 BCE to 2686 BCE has a stable average length of rule for the Pharaohs (20) of around 17 years each, with seventeen of those Pharaohs bearing the name of Horus (Hours).

*The Old Kingdom period of 2686 BCE to 2181 BCE begins the period of recorded global disasters. The early portion of the Old Kingdom sees a continued stability in rule, as twenty-two Pharaohs rule from 2686 BCE to 2375 BCE with an average reign of around 15 years each. Unas (6th Dynasty) becomes Pharaoh in **2375 BCE**, and an obelisk is dedicated to the Pharaoh's Jubilee in **2350 BCE**. I believe the onset of global disasters from this Planetary Passover brings the life of Unas to an end in **2345 BCE**, as he does not have to see the disaster to come. Teti succeeds Unas as Pharaoh in **2344 BCE**, and is the Pharaoh during the historical Global Flood. Teti rules until **2333 BCE**, as his death sees an end to this period of stability, as the next 20 years sees 10 Pharaohs that ruled **2** years each, and the next 17 years sees 4 Pharaohs that average **4.25** years of rule each. These periods of instability in ruler ship are in direct relationship to disasters versus political infighting. These Global Flood disasters brings the Old Kingdom to a crashing end.*

One striking coincidence shows the cosmic disasters, and the resulting instability that was caused led to brevity in pharaonic rule that has nothing to do with poor leadership or infighting! There are no images or statues of any pharaoh from the 7th, 8th, and 9th Dynasties! This was in large part to the deaths of skilled artisans, masons, teachers, astronomers, and more. The post-Flood 7th and 8th Dynasties pharaohs averaged 2 and 4 years respectively for rule, as this underscores the disasters affecting the populace. Someone find me an image of a pharaoh from this period, as I have searched high and low!

*The 9th Dynasty begins the **1st** Intermediate period of **2181 BCE** to **2055 BCE**, and is a time of continuing disasters, as this is the noted time of the destructions of the cities of Sodom, Gomorrah, Bela, Zoar, and Admah. The Nile River stopped flowing for 100 years, and over 80% of Egyptian cities are abandoned due to the conditions that existed. The average length of rule during this time is very brief at around **8** months per Pharaoh (**14**) from **2181 BCE** to **2169 BCE**.*

*The Middle Kingdom of **2055 BCE** to **1650 BCE** brings a period of stability with average lengthy reigns (**12th Dynasty**) of **25** years each, with **4** Pharaohs with **Amen** in their names. The Hebrew people enter the land of Egypt during this period, as upheaval begins with the **13th Dynasty** averaging **3.05** years reign for **38** different Pharaohs.*

*The **14th Dynasty** begins the **2nd Intermediate** period as the turmoil in rule continues with **17** Pharaohs* *ruling for a **3.05** year average. The **16th Dynasty** average rises to **16.66 years**, but the **17th Dynasty** begins to bottom out for a **7** year average rule.*

*The New Kingdom sees a stability in reign of almost **20** years per Pharaoh during the **18th Dynasty**, which is quite amazing considering the disasters of this epoch. Many of the Pharaohs during this time takes names with **Amen, Amun**, or **Mosis** in their titles. The Planetary Passover disasters continue as we reach the time of the Biblical period of the Exodus. The world sees this time from **1500 BCE** to **1447 BCE** continue the decimation of the planet. The Islands of Santorini in Greece and Dwarka in India collapse into the seas respectively around **1500 BCE**. The Hebrew people leave the land of Egypt in **1487 BCE** and finally settle in the land of Canaan in **1447 BCE**. The Egyptian culture continued a downward spiral as they never attain the high technological levels once achieved.*

Egyptian Pharaoh's List

<u>Old Pharaoh's list compiled from</u>: Palermo Stone, Turin List, Royal Canon, Manetho's Aegyptica, Abydos Kings List, Karnak Tablet, South Saqqara Stone, and the South Saqqara Tablet.

<u>Legend Period</u>

Turin	Manetho	Symbolizes
Ptah	Hephaistos	Craftsmen/Creation
Ra	Helios	Sun
Shu	Sosis/Sothis	Air
Geb	Kronos	Earth
Osiris	Osiris	Afterlife
Set	Typhon	Chaos
Horus	Horus	War
Thoth		Knowledge
Ma'at		

Turin Kings List	Manetho
2nd Dynasty of Gods	Dynasty of Half Gods
3 Achu Dynasties	30 Kings from Memphis
Disciples of Horus Dynasty	10 Kings from Thinnis

<u>Archaic Period</u>: Early Dynastic - Lower Egypt

<u>Palermo Stone Pharaohs</u>: Hsekiu, Khayw, Tiu Thesh, Neheb, Wazner, Mekh, Double Falcon

Early Dynastic: Upper Egypt

Pharaoh	Reign
Scorpion I	3200 BCE
Iry Hor	3150 BCE
Narmer	3150 BCE
King Scorpion	3150 BCE
Ka	3100 BCE

First Dynasty: 3150 BCE to 2890 BCE

Pharaoh	Reign
Narmer	3150 BCE
Hor-Aha	3050 BCE
Djer	?
Djet	?
Memeith	?
Den	?
Anedjeb	?
Semerkhet	?
Qa' A	2916 - 2890 BCE

First and Second Interim: 3150 BCE to 2890 BCE

Pharaoh	Reign
Sneferka	2900 BCE
Horus Bird	2900 BCE

Second Dynasty: 2890 BCE to 2686 BCE

Pharaoh	Reign
Hotepsekhemwy	?
Raneb	?
Nynetjer	?
Weneg	?
Sened	?
Seth-Paribsen	? Pyramids aligned 2740 BCE
Sekhemib-Perenmaat	?
Khasekhemwy	?

Old Kingdom: 2688 BCE to 2181 BCE (3rd to 6th Dynasties)

Pyramid Building Period

Third Dynasty: 2686 BCE to 2613 BCE

Pharaoh	Reign	Notable
Djoser	2668 - 2649 BCE	Builds 1st pyramid
Sekhemkhet	2649 - 2643 BCE	
Sanakhte	2650 BCE	
*Khaba	2643 - 2647 BCE	Built unfinished pyramid
Huni	2637 - 2613 BCE	

Fourth Dynasty: 2613 BCE to 2498 BCE

Pharaoh	Reign	Notable
Sneferu	2613 - 2589 BCE	Medium/Bent Pyramid
Khufu	2589 - 2566 BCE	Great Pyramid of Giza
Djedefra	2566 - 2558 BCE	Sphinx / Abu Rawash
Khafra	2558 - 2532 BCE	Middle Giza Pyramid
Menkhaura	2532 - 2503 BCE	Small Giza Pyramid
Shepsekaf	2503 - 2498 BCE	El-Fara'un Mastaba
Djedefptah		

Pre-Disaster Period

<u>Fifth Dynasty</u>: 2498 BCE to 2345 BCE

Pharaoh	Reign	Notable
Userkaf	2498 - 2491 BCE	Saqqara Pyramid
Sahure	2490 - 2477 BCE	Abusir Pyramid
Neferire Kare KaKai	2477 - 2467 BCE	
Shepsekare Isi	2467 - 2460 BCE	
Neferefre	2460 - 2425 BCE	
Nyuserre Ini	2425 - 2422 BCE	
Menkauhor Kaiv	2422 - 2414 BCE	
Djedkare Isesi	2414 - 2375	

	BCE	
*Unas	2375 - 2345 BCE	Unas dies prior to the Great Flood planetary passover of 2344 BCE

Global decimation after 2344 BCE Venus Planetary Passover and post-flood period until the 2184 Passover disaster!

End of 1st Mayan World Sun Nahui Ocelotl (Jaguar Sun): Tezcatlipoca - only became half a Sun by legend, and sends jaguars (Sun spots & meteorites) to destroy the giants (Biblical relationship)

<u>Sixth Dynasty</u>: 2345 BCE to 2181 BCE

Pharaoh	Reign	Notable
*Teti	2345 BCE - 2333 BCE	Flood Pharaoh
Userkare	2333 - 2332 BCE	Usurper?
Mereyre Pepi I	2332 - 2283 BCE	@2300 BCE over 80% of Egyptians cities are abandoned, and the first deserts appear on the planet
Merenre Nemtyemsaf I	2283 - 2278 BCE	
*Neferkare Pepi II	2278 - 2184	2184 BCE - **Sodom**

disputed length of rule	BCE	**and Gomorrah Passover disasters**
Neferka	2200 - 2199 BCE	
Nefer	2197 - 2193 BCE	
Aba	2193 - 2178 BCE	
?		
Merenre Nemtyemsaf II	2184 BCE	**Sodom & Gomorrah**
Neitiqerty Siptah	2184 - 2181 BCE	**Sodom & Gomorrah**

11 Pharaohs reign for 161 years averaging 14.9 years each

The 2184 BCE **Planetary Passover disasters that are recorded Biblically, as the cautionary Sodom and Gomorrah tale occurs 160 years (3 - 50 year Jubilees) after the Great Flood event.**

End of 2nd Mayan World Sun Nahui Ehecatl (Wind Sun): Quetzalcoatl (Venus) destroys man because they are corrupt, and turns them in to monkeys before sending hurakan -hurricane -evil - overthrow winds of lore.

<u>First Intermediate</u>: 2181 BCE to 2060 BCE   EXTREME TURMOIL

Pharaoh	Reign	Notable
Neferkara I	?	NO IMAGE/DATE
Netjerkare	?	NO IMAGE/DATE

Menkare	?	NO IMAGE/DATE
Neferkare II	?	NO IMAGE/DATE
Neferkare III Nebi	?	NO IMAGE/DATE
Djedkara Shemai	?	NO IMAGE/DATE
Neferkare IV Khendu	?	NO IMAGE/DATE
Merehor	?	NO IMAGE/DATE
Neferkamin Seneferka	?	NO IMAGE/DATE
Nikara	?	NO IMAGE/DATE
Neferkare V Tereru	?	NO IMAGE/DATE
Neferhahor	?	NO IMAGE/DATE
Neferkare VI Pepyseneb	?	NO IMAGE/DATE
Neferkamin Anu	?	NO IMAGE/DATE
Qakare Ibi	2169 - 2167 BCE	
Neferkara II	2167 - 2163 BCE	
Neferkawhor Khuwhap	2163 - 2161 BCE	
Neferirara	2161 - 2160 BCE	

This period of continuing turmoil sees 18 Pharaohs in 121 years for an average reign of 6.7 years each. Who wants to be a Pharaoh? This is the Biblical time of Abraham, and 2160 BCE to 0 CE is one zodiac age!

End of 3rd Mayan World Sun Nahui Quiahuitl (Rain Sun) - Tezcatlipoca steals Tlaloc's wife Xochiquetzal, and in his grief he refuses to send rain, and Tlaloc made it rain fire burning away the Earth.

Ninth Dynasty: 2160 BCE to 2130 BCE

Continued turmoil - 9 Pharaohs in 30 years for an average reign of 3.33 years each.

Pharaoh	Reign	Notable
? Achlthoes?	2160 BCE?	
?	?	
Neferkare VII	?	
Khety	?	
Senenh/Setut	?	
?	?	
Mer (Ibre Khety)	?	
Shed	?	
H?	?	

Tenth Dynasty: 2130 BCE to 2040 BCE (**Stable period**)

Pharaoh	Reign	Notable
Meryhathor	2130 BCE - ?	
Neferkare VIII	?	
Wankare	?	
Merykare	?	

Eleventh Dynasty: 2134 BCE to 1991 BCE

Pharaoh	Reign	Notable
Mentuhotep I	?	
Sehertawy Intef I	2134 - 2117 BCE	
Wahankh Intef I	2117 - 2069 BCE	

Pharaoh	Reign	Notable
Nakhtepnefer Intef III	2069 - 2060 BCE	

Middle Kingdom: 2060 BCE to 1802 BCE

Eleventh Dynasty continues into the Middle Kingdom

Pharaoh	Reign	Notable
Nebheterre Mentuhotep	2046 - 1995 BCE	
Sankhkare Mentuhotep III	?	
Nebtawyre Mentuhotep IV	?	

Twelfth Dynasty: 1991 BCE to 1802 BCE

Pharaoh	Reign	Notable
Sehetepi'bre Amenemhat I	1991 - 1962 BCE	
Kheperkare Sensuret I	1971 - 1926 BCE	
Nubkaure Amencmhat II	1929 - 1895 BCE	The Biblical Children of Israel enter Egypt @ 1917 BCE under Amenemhat II
Khakheperre Sensuret II	1897 - 1878 BCE	
Khakaure Sensuret II	1878 - 1860 BCE	
Nimaatre	1860 - 1815 BCE	

Amenemhat III		
Maakherure *Amenemhat IV*	1815 - 1807 BCE	
Sobekkare Sobekneferu	1807 - 1802 BCE	

<u>Second Intermediate Period</u>: 1802 BCE to 1550 BCE

<u>Thirteenth Dynasty</u>: 1802 BCE to 1649 BCE

Sekhemrekhutawy Sobekhotep / Wegaf?

Pharaoh	Reign	Notable
Sekhemrekhutawy Sobekhotep / Wegaf?	1802 - 1799 BCE	
Sekhemkare *Amen*emhat V	1799 - 1796 BCE	
*Amen*emhat ?	1795 - 1792 BCE	
Sehetepre	1792 - 1790 BCE	
Iufni	1788 or 1790 BCE	
Seankhhibre *Amen*emhat VI	?	
Semankare	?	
Sehetepre	?	
Sewadjkare	?	
Nedjemibre	? @ 7months	
*Kha*ankhre Sobekhotep I	1780 - 1777 BCE	
Renseneb	1777 BCE @ 4 months	
Awybre Hor I	1777 - 1775 BCE @ 18 months	
Sedjefakare	1769 - 1766 BCE	

Sekhermae Khutawy Sobekhotep II	1767 BCE	
Khendjer	1765 BCE	May have been the first Semitic Pharaoh?
Imyremeshaw	1759 BCE	
Sehetepkare Intef	?	
Seth Meribel	?	
Sekhemresewadjtawy Sobekhotep III	1755 - 1751 BCE	
Khasekhemre Neferhotep I	1751 - 1740 BCE	
Khaneferre Sobekhotep IV	1740 - 1730 BCE	
Merhotepre Sobekhotep V	1730 BCE	
Khahotepre Sobekhotep VI	1725 BCE	
Wahibre Ibiau	1725 - 1714 BCE or 1712 - 1701 BCE	
Merneferre Ay	1701 - 1677 BCE	longest reign of dynasty
Merhotepre Ini	1677 - 1675 BCE	
Sankhenre Sewadjtu	1675 - 1672 BCE	
Mersekhemre Ined	1672 - 1669 BCE	
Sewadjkare Hori	1669 - 1664 BCE	
Merkawre Sobekhotep VII	1664 - 1663 BCE	
Seven Kings (7)	?	
Mer(?)re	?	
Merkheperre	?	
Merkare	?	

Pharaoh	Reign	Notable
?	?	
Sewedjare Mentuhotep V	?	
(?) Mosre	?	
Ibi (?) maatre	?	
Hor(?) (?)webenre	?	
Se(?)kare	?	
Seheqenre Sankhptahi	?	
(?)re	?	
Se(?)enre	1649 BCE	
Dedumose I	@ 1654 BCE	
Dedumose II	?	
Senebriu	?	
Senaaib	?	

This continuing period of ecological turmoil sees 54 Pharaohs reign for 153 years, and average 2.83 years each. This turmoil is not solely based on political infighting, as the infighting is based on the condition of the land!

<u>Fourteenth Dynasty</u>: 1705 BCE to 1690 BCE

Pharaoh	Reign	Notable
Nehesy	@ 1705 BCE	
Khakherewre	?	
Nebafawre	@ 1704 BCE	
Sehebre	?	
Merdjefare	@ 1699 BCE	
Sewadjkare	?	
Needjefare	@ 1694 BCE	

Webenre	?	
?	?	
Djefare	?	
Webenre	@ 1690 BCE	
Sheshi	?	
Yakubher	?	

Fifteenth Dynasty: 1674 BCE to 1535 BCE

Pharaoh	Reign	Notable
Salitis	?	
Sakir-Har	?	
Khyan	30-40 years	
Apepi	40 years +	
Khamudi	155 5 - 1544 BCE	

Sixteenth Dynasty: Local dynasty that lasted about 100 years and saw 15 Pharaohs rule for an average of 6.66 years each.

Seventeenth Dynasty: 1650 BCE to 1550 BCE

Pharaoh	Reign	Notable
Sekhemrewahkhaw Rahotep	1620 BCE	
Sekhemre Wadjkhaw Sobekemsaf I	?	
Sekhemre Shedtawy Sobekemsaf	?	
Sekhemre -	?	

Wepmaat Intef		
Nubkheperre Intef	?	
Sekhemre - Heruhirmaat Intef	?	
Senakhtenre Ahmose	1558 BCE	
Seqenenre Tao	1558 - 1554 BCE	Died in a battle with the Hyksos
Kamose/Khamose	1554 - 1549 BCE	

<u>New Kingdom</u>: 1550 BCE to 1077 BCE

<u>Eighteenth Dynasty</u>: 1550 BCE to 1292 BCE

Pharaoh	Reign	Notable
Nebpehtire Ahmose I, Ahmosis I	1550 - 1525 BCE or 1570 - 1544 BCE	
Djeserkare Amenhotep I	1541 - 1520 BCE	
Aakheperkare Thutmose I	1520 - 1492 BCE	
Aakheperenre Thutmose II	1492 - 1479 BCE	Pharaoh of the Exodus disasters in 1487 BCE
Maatkare Hatshepsut	1479 - 1458 BCE	Dies before the Exodus Children enter the Holy Land in 1447 BCE
Menkheperre Thutmose III	1479 - 1425 BCE	Post-disaster co-Pharoah with Hatshepsut
Aakheperrure Amenhotep II	1425 - 1400 BCE	
Menkheperure	1400 - 1390 BCE	

Thutmose IV		
Nebmaatre Amenhotep III The Magnificent King	1390 - 1352 BCE	
Neiferkheperure - waenre Amenhotep IV/Akhenaten	1352 - 1334 BCE	
Ankhkheperure Smenkhkare	1334 - 1333 BCE	
Nebkheperure Tutankhaten / Tutankhamun	1333 - 1324 BCE	
Kheperkheperure Ay	1324 - 1320 BCE	
Djeserkheperure - setpenre Horemheb	1320 - 1292 BCE	

The periodic disasters continue in the Eighteenth Dynasty of the New Kingdom, as the Biblically related disasters of the Exodus play themselves out. The planetary Passover event in 1487 BCE brings more upheaval to the land, society, and rulership, that causes the Children of Israel to wander the desert for 40 years until 1447 BCE.

End of 4th Mayan World Sun: Water Sun - Nahui Atl

Calchiuhtlicue becomes the Sun God of the World, as Tezcatlipoca (old Sun) and Quetzalcoatl (Venus) continue their feud as they strike down the Sun (Calchiuhtlicue) and flood the Earth. The Earth is cast into darkness as

Quetzalcoatl (Venus) descends (below the horizon for 3 days) to the underworld (Innana, Dummuzi, Christ) to bring up the bones of the dead (graves overturned). The descent of Quetzalcoatl (Venus) into the underworld was a concept readily understood by the Aztecs when Spanish Conquistadors and Missionaries arrived. The 4th World comes to an end.

The fifth sun - Earthquake Sun - Tecuciztecatl

Tecuciztecatl offers to be the new Sun God, but Nanahuatzin is chosen in his place. Tecuciztecatl lacks the courage to jump into the fire as Nanahuatzin jumps in followed by an eagle (Aquila constellation) and a jaguar (Sun spots). The overwhelming brightness caused the Gods to throw a rabbit (Hare-Lepus constellation) over Tecuciztecatl, dimming his glow, causing him to become the Moon. This covering of the Moon is a common theme in sacred texts and documents around the globe. The 5th Sun is the time that we currently live in.

Nineteenth Dynasty: 1292 BCE to 1186 BCE

8 Pharaohs rule for a total of 106 years averaging a stable 13.25 years each, with the Pharaoh known as Ramesses II the Great being the most recognizable.

Twentieth Dynasty: 1190 BCE to 1077 BCE

10 Pharaohs reign for 143 years averaging for 14.3 years each. The Pharaoh Usermaatre-meryamun Ramesses III fights a noted battle against the Sea Peoples in 1175 BCE.

Third Intermediate Period: 1077 BCE to 732 BCE

Twenty-First Dynasty: 1069 to 943 BCE

This period saw an end to Egyptian rule in the land as Libyans, Libu, Romans, Nubians, Persians, and Macedonians rule at various times. The technology of the society began a downward spiral after the Great Flood event, never attaining the high levels of pyramid construction.

Star Representations and Alignments

Pyramid	Star Represented	Star Name
Khufu - Great Pyramid	Alnitak	Delta Orionis
Kephres Pyramid	Alnilam	Epsilon Orionis
Menkeres Pyramid	Mintaka	Delta Orionis
Sphinx	Rigel	Beta Orionis
Abu Rawash Pyramid	Saiph	Kappa Orionis
Red Pyramid	Aldebaran	Alpha Tauri
White/Black Pyramid		Epsilon Tauri
El-Anazan Pyramid	Bellatrix	Gamma Orionis

Star Shaft Alignment	Star
Southern Star Shaft	Sirius
Southern Star Shaft	Alnitak
Northern Star Shaft	Thuban * Pole Star 2787 BCE - pyramids aligned in 2740 BCE
Northern Star Shaft	Kochab

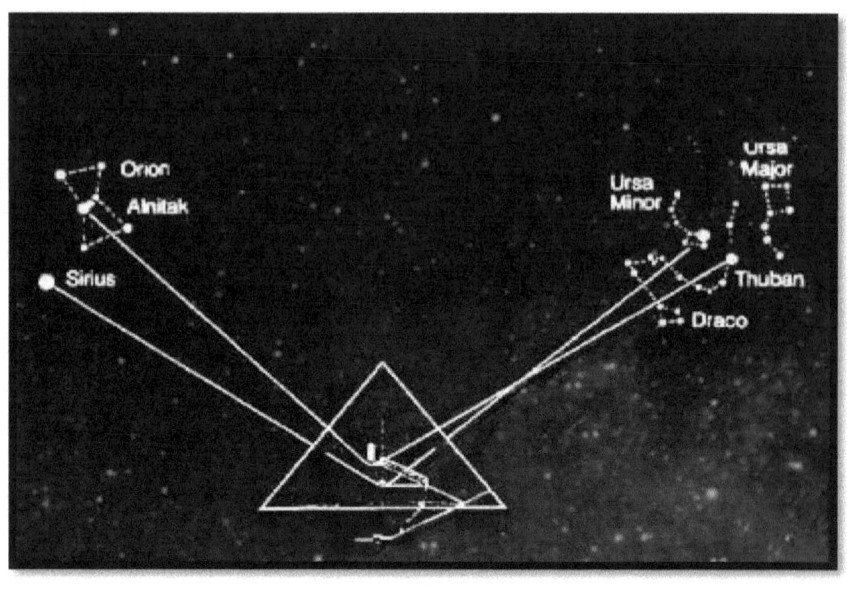

Pyramid Facts

The Great Pyramid is estimated to have around 2,300,000 stone blocks that weigh from 2 to 30 tons each, with some weighing over 50 tons.

The pyramids of Khufu (Great), Menkaure, and Kephres, are precisely aligned with the Belt Stars in Orion; Alnitak, Mintaka, and Alnilam. The three pyramid's size accurately reflects the stars size in relation to the cosmic heavens on Earth.

The Queen's chamber interior temperature is a constant 68°s which equals the average temperature of the Egypt.

*Curious: Avg. temp. 68° - 102° high - 34° low - below 33°f is freezing.

The Great Pyramid was originally covered in 144,000 outer casing stones of highly polished limestone that weighed up to 15 tons each.

  144,000 days ÷ 8 concave sides = 18,000 days per side
  18,000 days ÷ 360 days = 50 years per side
  Pharaoh's Jubilee was every 50 years
  50 years X 8 sides = 400 years - Tropical Year

  * 144,000 days = 1 Mayan Baktun = 394.26 years
    1 Tropical/Solar Year = 394.26 years

  Rev. 7:4; there will be 144,000 sealed for each of the 12 Tribes of Israel (12 zodiac signs) or 144,000 ÷ 12

=   12,000 days each sign ÷ 360 = 33.33 (never fractions or   decimals in the Egyptian sums).

The mortar used is of a known composition, but cannot be reproduced today. The mortar is stronger than the stone itself!

The Great Pyramid is aligned to True North with only 3/60th a degree of error, and this is only due to precession (?)

The base of the Great Pyramid covers 592,000ft.2, and this number is a curious constant: 25,920 = Zodiac Age, and 2,592,000 equals the mean astronomical distance of .0026au of the Earth and Moon.

The Great Pyramid with the capstone attached and lit could be seen all the way to the mountains in Jerusalem.

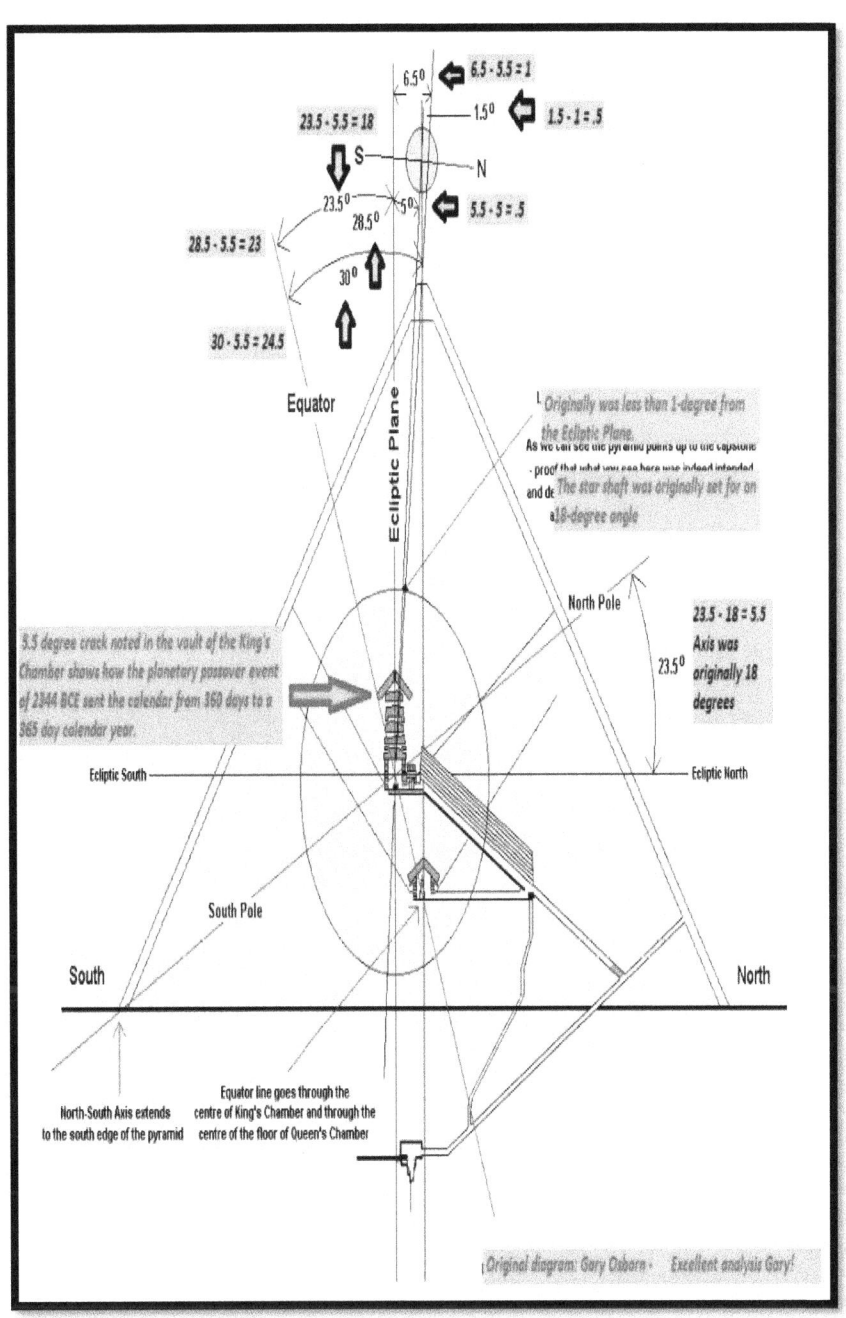

The Sphinx is a dog!

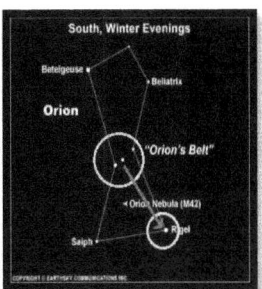
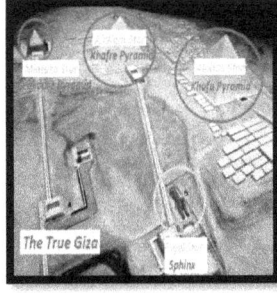

I do not feel that the Sphinx is the representation of a lion, but it is the body of a dog! Of course this flies in the face of scholarly convention, but I have several key points to my credit. The Sphinx sits in a key position facing East toward the rising Sun, and is angled in between the three pyramid complex. We know that the three pyramids represent the Belt Stars in Orion, and based on this the Sphinx would logically represent the star Rigel. The stars in Orion are most prominent in the southern hemisphere during the months of January to March (324° 11/24 to 72° 3/13) signaling the Summer Season! Rigel is known as the Foot or Ankle star of Orion in astronomy and sits at the feet of the pyramids.

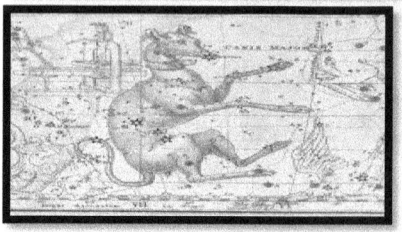

One can easily notice how slender the body, paws, and legs of the sphinx are, and that they resemble those of a dog in build more so. The big tell is the tail! A lion is the only big cat that has a tuft of hair at the tip of its tail, that is used as a weapon. The Sphinx never had this definitive tuft, and this would not have been left out by the builders, engineers, architects, artists, and masons that built these structures that still cannot be reproduced to this day. The current refurbishing efforts clearly shows no tuft was ever intended! There are lion representations created by the Egyptians that clearly show the tail with tuft of hair. I have come to the conclusion that the Sphinx was the representation of a dog, as the dog days of summer were signaled with the appearance of Rigel at that degree.

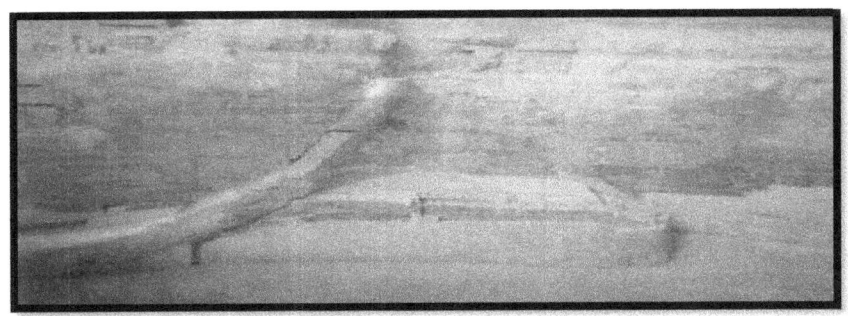

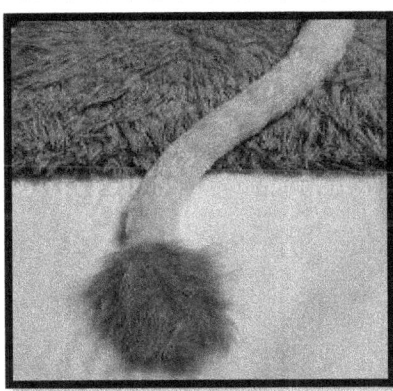

Giza Power Plant Live

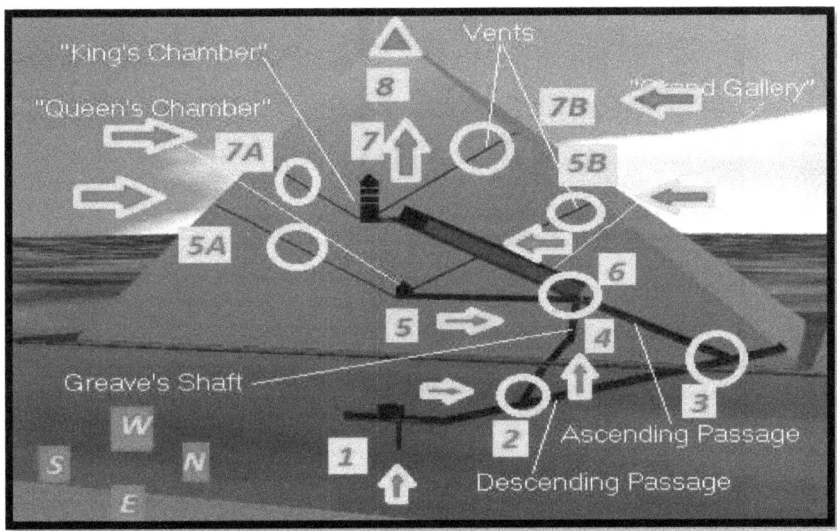

It is supremely clear that the Great Pyramid of Giza was built for a utilitarian reason and was never a tomb for a Pharaoh or anyone else. The Giza complex was primarily concerned with electrical production, distribution, and timekeeping. The exact process by which electricity was produced is still unknown at this point, but the tell tale signs of electrical activity are apparent. The staining and patina of the vaults (capacitors) shows electrical arcing and residue. The vertical shafts that were aligned to the stars had blocks placed in their path to help the electrical reaction stop. The water from the aquifer(1) was ionized as it ascended the descending passage (2), mixing with the particles from the Queen's Chamber (5) heading up to the Grand Gallery (6), where it met the King's Chamber mixture (7) and was disseminated somehow through the apex (8).

King's Chamber patina *Grand Gallery patina*

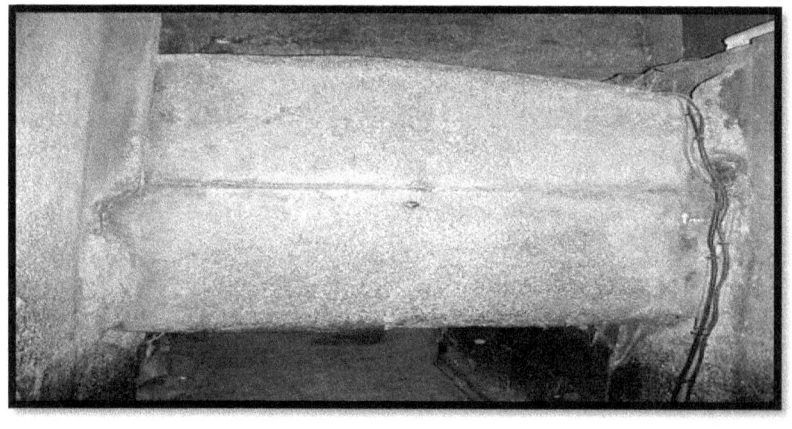

Granite weathering from the reaction process

Queen's Chamber star shaft block   Kings Chamber star shaft

Coral Castle Capacitor

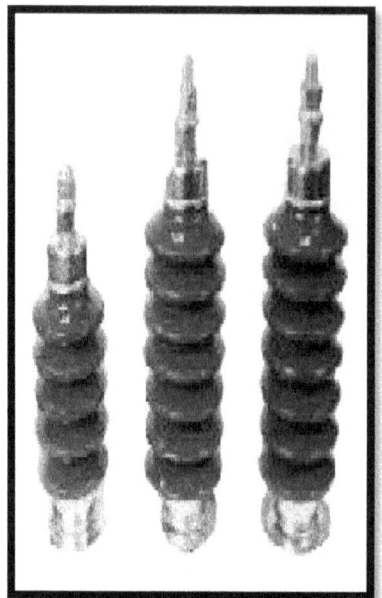

Ceramic capacitor bushings

Kings Chamber capacitor

World Culture of Squaring the Heavens

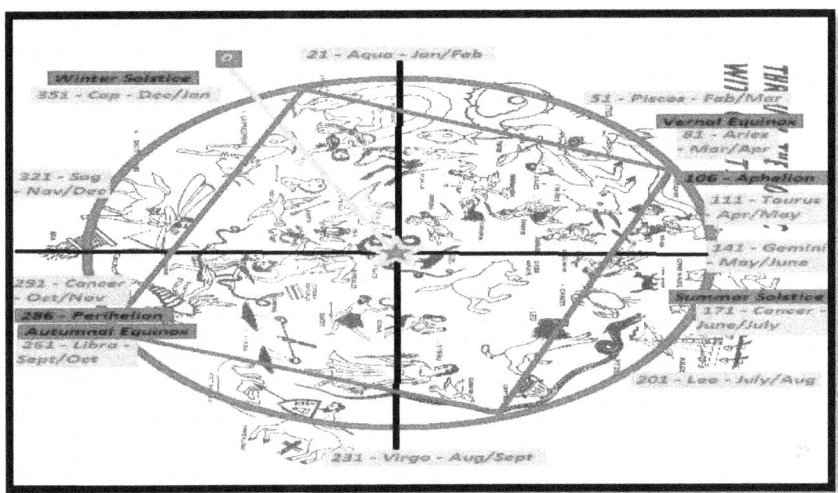

The 48 constellations of Ptolemy from @ 150 CE shows Aquarius at the 90° point, but the Dendera constellation disc has it at about the 30° point.

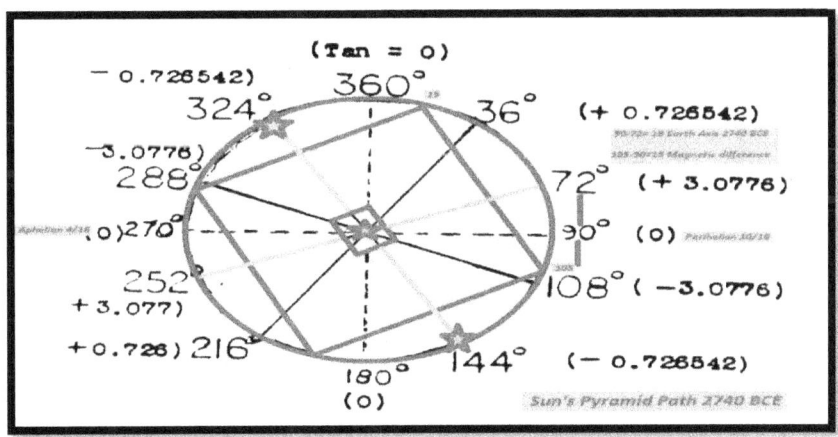

This diagram of solar cycles pre 2344 BCE gives us the true position of the Sun's path.

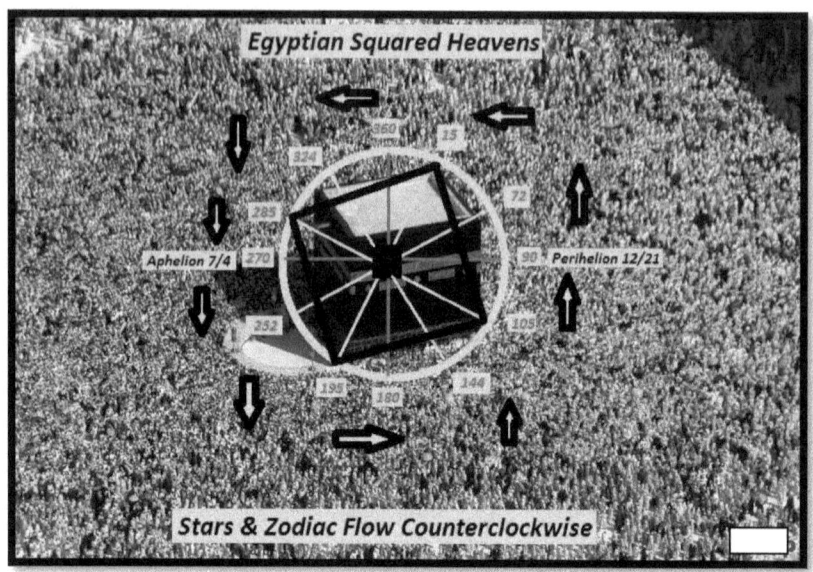

The Islamic faithful circumambulate the Kabaa (Pharaoh Khaba 2643 BCE) in the representation of squaring the heavens in Egyptian symbolism.

 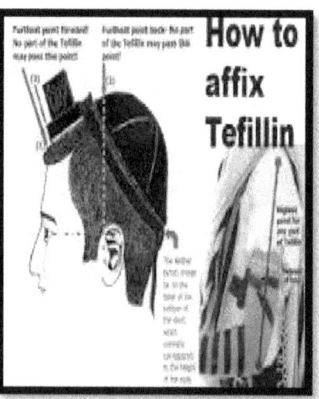

The male Jewish faithful wear the squared cube known as a tefillin. All of these parallels with Egypt, Islam, and Judaism, shows a shared but splintered understanding.

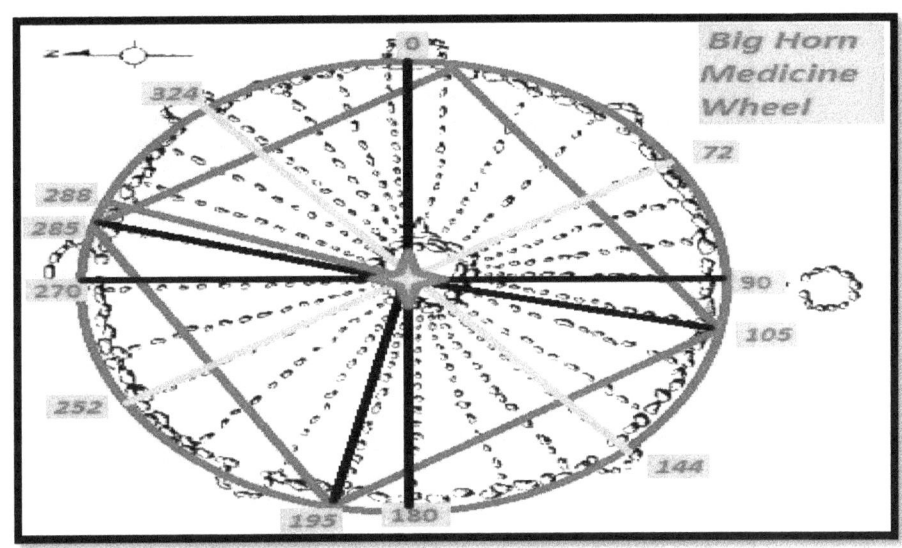

Big Horn Medicine Wheel - Wyoming, US

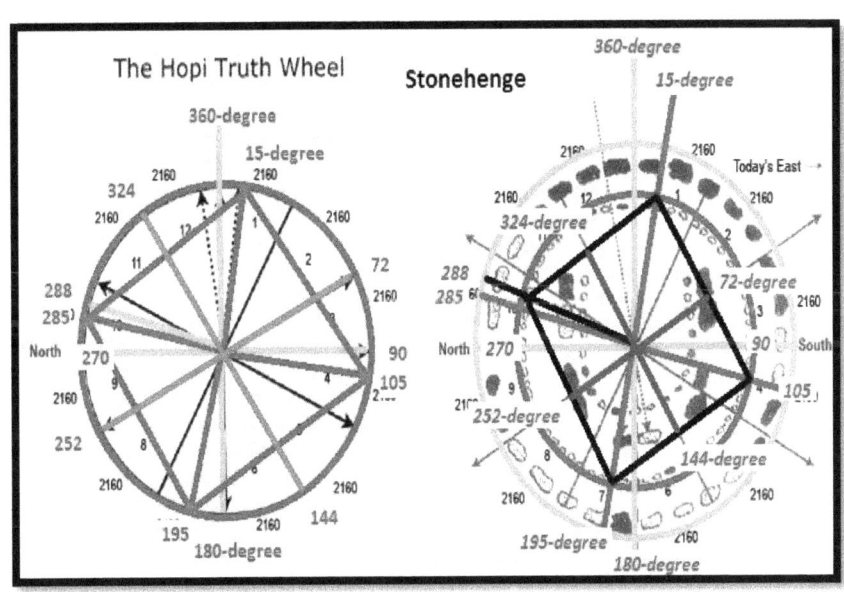

Hopi Truth Wheel & Stonehenge are off of Eastward orientation

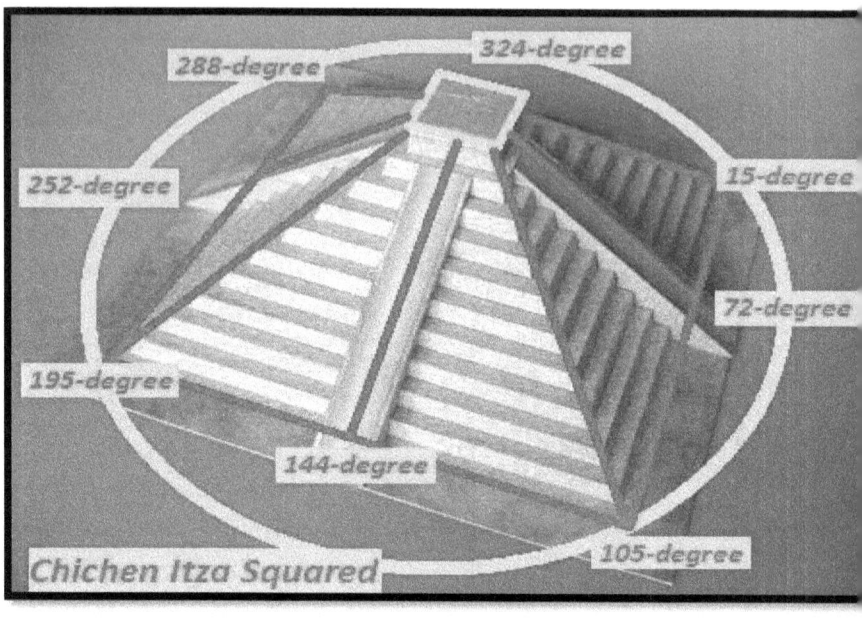

Chichen Itza was built @ 600 CE to track the equinoxes/solstices

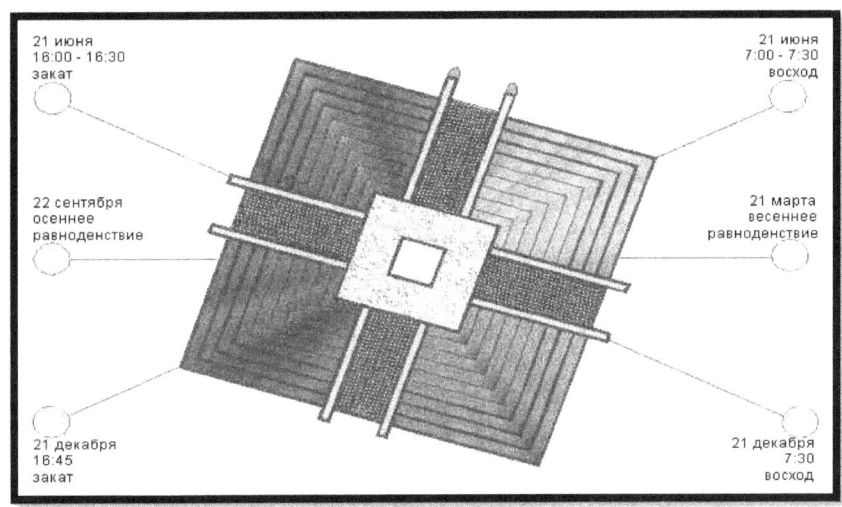

Pyramid of Kukulkan

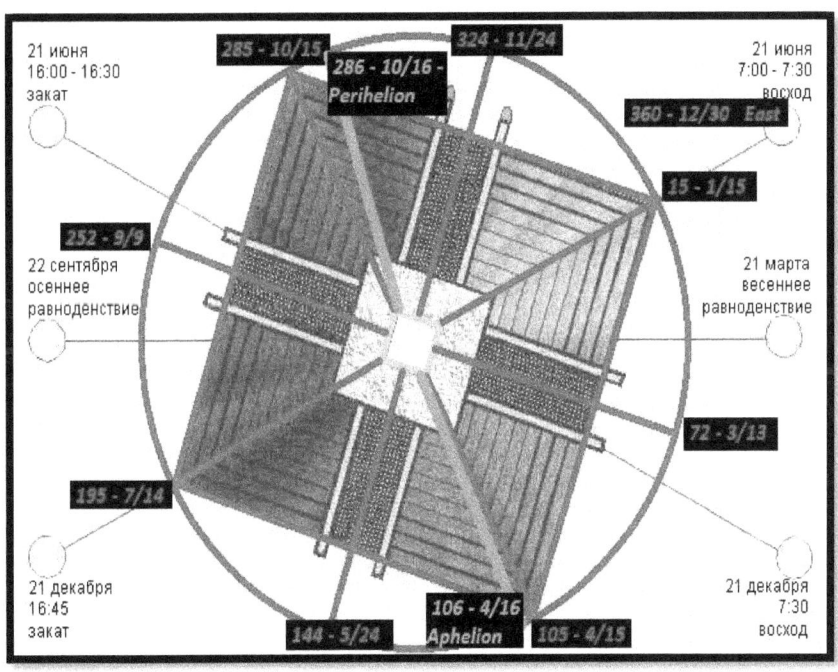

Monitoring the Solar Phases

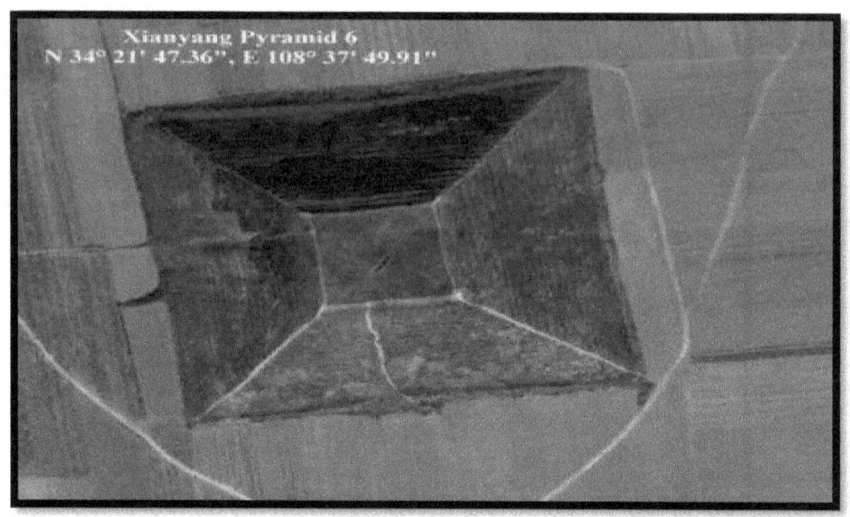

Xianyang Solar Pyramid - China

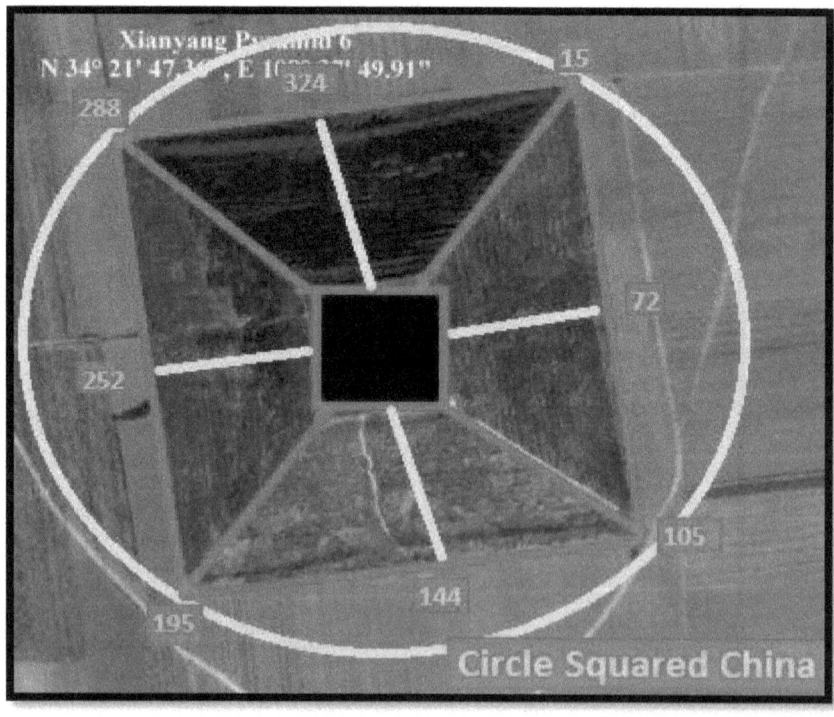

The Dendera Disc has been dated from 2500 BCE to 150 CE. This zodiac shows Aquarius at 30°s off from the present 90° point.

Global Pyramid Culture

Bosnia Squared

Candi Sukhuh - Indonesia

China Squared

Inner Mongolia Squared

Italy Squared

Khymer Pyramid Squared - Cambodia

Merope Pyramids - Sudan
(was the missing capstone an electrical receiver?)

Conclusions

During the course of research on this project, it has become crystal clear as to the superiority in astronomy, masonry, engineering, architecture, mathematics, and more, that the ancient Egyptians possessed even over modern day man. The pyramids at Giza and other sites exhibit what we have come to call Pi, Phi, Fibonacci numbers, and the Golden Ratio. These modern systems of math have been attributed to others, as the pyramids were built two-thousand years before people like Pythagoras (570 BCE - 495 BCE) were born. The knowledge base of the Egyptians were assimilated by other younger cultures such as the Greeks, Hebrews, Romans, and several others through the Egyptian libraries. It is clear that the Greek and Roman building boom came after their contact with the Egyptian culture.

The heavens and earth were skillfully mapped for being the 360-degree orbs that they are from our perspective. The 360-degrees in the circle was assigned to the 360-day calendar year, as the circle is a symbol of wholeness or completion as well. The Great Pyramid at Giza served as an electricity generating power plant and as a solar calendar capable of tracking days, weeks, months, years, Jubilees, and eons of great length.

The Great Pyramid is the most accurately aligned to true north building in the world 4700 years later. When we *square the Great Pyramid with a super-scribed circle with degree, day, season, or year number assignment, we can get a better understanding of the simple genius in the system. The 8 concave sides of the 4 apparent sides are designated as followed:*

North face: 195° (7/15) Winter to 252° (9/12) Spring to 285° (10/15) - Winter 57 days + Spring 33 days

East face: 285° (10/15) Spring to 324° (11/24) Summer to 15° (1/15) - Spring 39 days + Summer 51 days

South face: 15° (1/15) Summer to 72° (3/13) Fall to 105° (4/15) - Summer 57 days + Fall 33 days

West face: 105° (4/15) Fall to 144° (5/24) Winter to 195° (7/15) Fall 39 days + Winter 51 days

The Vernal Equinox occurs at the 81° (3/21) point and the Autumnal Equinox occurs at the 261° (9/21) mark. The aphelion takes place at 106° (4/16), while the perihelion occurs at the 286° (10/16) mark. The Summer Solstice is at 171° (6/21) degree, and the Winter Solstice point is the 351° (12/21) mark. When we remember that the Great Pyramid is not four-sided, but has eight concave sides with 144,000 exterior casing stones we are able to discern the use. The 144,000 casing stones represent a tropical/solar year of 144,000 days that gives us 400 years in 360 day calendar years. The 400 years break down to 50 year Jubilees for each of the eight noted pyramid sides.

The holidays that we celebrate as a religion or society have been predicated on these solar understandings that have been misunderstood. The perihelion in 2012 occurs at the 3° mark as January 3rd, and is celebrated as New Years Days and the Solemnity of Mary. The Vernal Equinox at 81° (3/21) is celebrated as Vernal Equinox Day, March Equinox Day, and Ostara. The all important Passover point at 106° (4/16) signals the aphelion point in pre 2344 BCE, and is celebrated as (4/15 - 105°) Passover Day, Tax Day, and the Bengali New Year. The Summer Solstice of 2012 occurs on 111° (4/21) as Earth Day. The 139° (5/19 Pentecost) ends the celebration of Passover season 91° (4/1 April Fools, Catholic Easter, First Day of Passover).

The pre 2344 BCE Summer Solstice occurred on 6/21 at 171°, and is celebrated as Litha and June Solstice Day. The current aphelion of 185° occurs on July 4th and is celebrated as the 4th of July or Independence Day! The former Autumnal Equinox occurred at 261° now passes at the 264° mark on 9/21, and is remembered as the Feast of Mabon, Mid-Autumn Festival, September Equinox Day, and Autumnal Equinox Day.

The former perihelion passed at the 286° point on 10/16 and is celebrated as Hajj and Eid-al-adha in the Islamic faith. The former Winter Solstice was 351° as 12/21, and takes place at the 355° point as 12/21 in 2012, and this is celebrated as Christmas, Yule Day, and Winter Solstice Day.

The current religious and political holidays have been inextricably assigned to these solar points, as symbols of divinity and societal celebrations. The mind that finds itself still contained with the box will find reasons to explain away this information that runs counter to their faith or belief systems. The founding fathers in America delayed the signing of the Declaration of Independence until July 4th, a day that they considered more auspicious to their end. The death of Jesus Christ and Tammuz are celebrated on December 21st the shortest day of the year celebrating the death of the sun. The sun is reborn on the 25th and is remembered as Christmas in the Christian and Catholic faiths.

The former path of the sun prior to the planetary Passover event that occurred in 2344 BCE was different than that of today in 2012 CE. The southern hemisphere's Summer Solstice took place on 3/13 at the 72° point, the Autumnal Equinox passed on 5/24 at the 144° point, the Winter Solstice passed at the 252° mark on 9/12, and the Vernal Equinox took place on 11/24 at the 324° point.

The aphelion occurred on 4/16 at the 106° point, while the perihelion passed on 10/16 at the 286° mark in 2344 BCE The aphelion in 2012 CE takes place on 1/3 at the 3° mark, and the perihelion passes on 7/4 at the 185° point. There is a notable difference between the two aphelions of 79 days, and a difference of 77 days between the two perihelions. The mean difference would be 78 days, as this was not a result of precession and took place *overnight due to the planetary Passover. The passage of Venus when we were at our furthest point from the sun pushed the earth down 5.5°s and forward 78 days. This event added 5.5 days to the earth's calendar by slowing down the rotation from an 18-hour day to a 24-hour day, and altering the axis from 18°s to 23.5°s (+5.5°s). Overnight the world (Southern hemisphere) went from fall to winter, being ill prepared for the event, and this search to reckon the seasons is remembered in Biblical literature.*

*The 5.5°s shows the damage that was inflicted on the planet on Friday - April 13, 2344 BCE. The earth may have still been the continent of Pangea at this point? The Great Pyramid sits at the navel of the planet and may have caused the parabolic trajectory to take this path due to the electromagnetic effects. I have outlined in other books the Disaster-Miracle Path (DMP) of Biblical scripture that runs from the Black Sea (**Black Horse** - Great Flood 2344 BCE) to the Sea of Galilee (**White Horse** - Miracles of Christ 0 CE), then onto the Dead Sea (**Pale Horse** - Sodom & Gomorrah 2184 BCE), and ends in the Red Sea (Red Horse - Exodus Disasters 1487 BCE) in an almost straight line. I believe the location and function of the Great Pyramid made this parabolic trajectory even more deadly at times.*

The approach of Venus took a northeast direction, and this is represented in the 5° crack in the massive capacitor (relieving chamber) beams in the King's Chamber. The improperly termed King's Chamber was never meant to be a repository for anyone's deceased body, but was an integral area for the electrical reaction to take place. The capacitor beams estimated weight range from;

60m top slab - 5th - 132,277lbs
57.5m 4th slab - 126,765.8lbs.
54.6m 3rd slab - 120,372.4lbs.
51.9m 2nd slab - 114,419.9lbs.
48.8m bottom slab - 1st - 107,585.6lbs.
43m - base stone - 94,798.8lbs.

With an increase in the weight of the slabs, we see a step up in voltage capacitance from the lowest beam at 48.8m to the highest at 60m at the top.

The 5° crack in the King's Chamber also shows a correlation to the angle of the earth and moon. The moon sits at a 5° declination to the earth, which supports the capture theory, as the moon would have originally sat at the equator or 0° level.

The star alignment date of the Giza Pyramids of 2740 BCE relates directly to the Great Flood of 2344 BCE, as it is 5.5 celestial portions (72yrs per portion) or 396 years to that event.

2740 BCE
-2344 BCE
396 years or 5.5 celestial portions!!!

This 5.5 change is mirrored as well in the change of 400 years or 144,000 days being counted as a tropical or solar year. The tropical/solar year now is 394.52 a difference of 5.48 years.

The recorded memory of mankind versus the actual events has been very spotty. Disasters have been related millennia after the event, and typically with a religious fervor claiming deity to these disasters. Mankind reawakens around 4000 BCE after another series of catastrophes that periodically came from the heavens. Sumerian culture remembered this reawakening in fables as Adapa and Titi. The Torah and Bible reworks this tale as Adam and Eve Biblically, some 2,500 years or more later.

The Egyptian, Hindu, and Mayan civilizations start their calendar systems around 3100 BCE each. The Global Great Flood of 2344 BCE brings a fond remembrance for the Age of Taurus that ends in 2309 BCE, and a lack of recognition for the new Age of Aries that entered in 2308 BCE. Scientists have recently concluded that the first deserts did not appear on Earth until 2300 BCE. This dovetails nicely into the timelines that my research had previously established. I have consistently questioned why we have deserts? How does sand exactly form? Why do we have salt water? A Planetary Passover by another charged body would definitely bring those resulting land and climate changes.

The period of 2184 BCE brings continued disasters as the Nile ceases to flow for 100 years, 80% of Egyptian cities are abandoned, and the Biblically related disasters of Sodom, Gomorrah, Bela, Admah, and Zoar, take place *during this time. This is the time period of Abram in Hebrew, Islamic, and Christian faiths.*

There comes a period of recovery, until another series of disasters takes place in 1500 BCE to 1447 BCE, as Thera in Greece and Dwarka in India sink into the sea respectively, and the Exodus disasters unfold in Egypt. A new glassy material that is discovered by travelers in the newly formed deserts is linked to the God's in the sky that formed it. These pieces of plasma blasted sand became a favorite of Pharaohs and the ruling elite to wear as ceremonial jewelry. We call this new jewel Libyan Desert Glass, and this type of glass can only be recreated through a nuclear explosion.

These periodic disasters of remembered times show direct parallels with the length of reign for the Egyptian Pharaohs and the corresponding kingdoms. The term pharaoh is Hebrew for the Egyptian term "per aa." The Early Dynastic period of 3100 BCE to 2686 BCE begins recorded Egyptian history, and has a very stable average length of rule for the Pharaohs (20) of around 17 years each, with seventeen of those Pharaohs bearing the name of Horus (Hours).

The Old Kingdom period of 2686 BCE to 2181 BCE begins the period of recorded global disasters. The early portion of the Old Kingdom has a continued stability in rule, as twenty-two Pharaohs rule from 2686 BCE to 2375 BCE with an average reign of around 15 years each.

Unas (6th Dynasty) becomes Pharaoh in 2375 BCE, and an obelisk is dedicated for the Pharaoh's Jubilee in 2350 BCE. The Jubilee is for the 50th year, versus 25 years for the rule of Unas. I believe the onset of global disasters from this Planetary Passover brings the life of Unas to an end in 2345 BCE, as he is taken and does not have to see the disaster to come. Teti succeeds Unas as Pharaoh in 2344 BCE, and is the Pharaoh during the historic Global Flood. Teti rules until 2333 BCE, as his death sees an end to this period of stability, as the next 20 years sees 10 Pharaohs that ruled 2 years each, and the next 17 years sees 4 Pharaohs that average 4.25 years of rule each. These periods of instability in ruler ship are in direct relationship to disasters versus political infighting. These Global Flood disasters brings the Old Kingdom to a crashing end, as no images are to be found of these Pharaohs.

The 1st Intermediate period of 2181 BCE to 2055 BCE is a time of continuing disasters, as this is the noted time of the destructions of the cities of Sodom, Gomorrah, Bela, Zoar, and Admah in 2184 BCE. The Nile River stopped flowing for 100 years, with over 80% of Egyptian cities being abandoned due to the conditions that existed. The average length of rule during this time is very brief at around 8 months per Pharaoh (14) from 2181 BCE to 2169 BCE. The brevity in leadership is not due to political turmoil or wars, but directly to the natural (unnatural) deaths of leadership.

The Middle Kingdom of 2055 BCE to 1650 BCE brings a period of stability with average lengthy reigns (12th Dynasty) of 25 years each, with 4 Pharaohs using Amen in their names. The Hebrew people enter the land of Egypt during this period, as upheaval begins with the 13th Dynasty averaging 3.05 years reign for 38 different Pharaohs.

The 14th Dynasty begins the 2nd Intermediate period as the turmoil in rule continues with 17 Pharaohs ruling for a 3.05 year average. The 16th Dynasty average rises to 16.66 years, but the 17th Dynasty begins to bottom out for a 7 year average rule.

The New Kingdom sees a stability in reign of almost 20 years per Pharaoh during the 18th Dynasty, which is quite amazing considering the disasters of this epoch. Many of the Pharaohs during this time takes names with Amen, Amun, or Mosis in their titles. The Planetary Passover disasters continue as we reach the time of the Biblical period of the Exodus. The world sees this time from 1500 BCE to 1447 BCE continue the decimation of the planet. The Islands of Santorini in Greece and Dwarka in India collapse into the seas respectively around 1500 BCE. The Hebrew people leave the land of Egypt in 1487 BCE and finally settle in the land of Canaan in 1447 BCE. The Egyptian culture continued a downward spiral as they never attain the high technological levels once achieved.

The pyramid complex at Giza was designed to reflect the stars of Orion's Belt as Khufu's pyramid represents Alnitak (Delta Orionis), Kephres pyramid is Alnilam (Epsilon Orionis), Menkeres pyramid is Mintaka (Delta Orionis), and the mistaken Sphinx represents Rigel (Beta Orionis). The pyramids up and down the Nile river represents heaven on earth in symbolism and function, as the Abu Rawash pyramid reflects Saiph (Kappa Orionis), the Red Pyramid is Aldebaran (Alpha Tauri), the White/Black pyramid is Epsilon Tauri, and the El-Anazan pyramid represents Bellatrix (Gamma Orionis).

These masters of so many disciplines clearly depicted the stars in their proper position and size for their utilization. These builders were not preoccupied with death, but were fixated with the creation and perpetuation of life.

The Sphinx is a dog! This view flies in the face of convention (con), as the prevailing opinion regards it is a lion in representation of the Leo constellation. A few points support my theory on this as Rigel is the foot or ankle star in the Orion constellation, and sits at an angle to the middle belt star Alnilam (Epsilon Orionis) as does the Sphinx. This representation makes more sense as Rigel signals the dog days of summer with its appearance in the southern hemisphere in January.

One need only look at the slender elongated front and hind legs, and slender body to see that of a dog versus a lion. The paws are clearly more canine in appearance even with the uneven restoration effort. The tail of the Sphinx is missing the obvious tuft of hair that all adult male *lions have, and is often used as a whipping weapon in battle. The master builders that represented all things in golden ratio would not have forgotten this distinctive feature.*

I have not been able to figure out the mechanism by which the Great Pyramid produced electricity, but the design of the pyramid hints at the use. The solar calendars use was secondary to the primary purpose of electrical production and distribution.

The pyramid complex faces east toward the rising sun on the 285° to 15° point, causing the aquifers under the Nile to ionize the water through the rocks into the subterranean chamber. These ionized particles head up through the ascending and descending passages. The particles from below then meet up with the electrical particles produced in the Queen's Chamber. This new mixture heads up the Grand Gallery to the King's Chamber where the final mix occurs, before heading up through the center axis to the capstone for distribution. How did all of these processes take place? That is the question that has befuddled mankind since the pyramids were rediscovered.

Our modern civilization believes that we are at the zenith of humanities understanding and abilities, but it is painfully obvious that we have taken some steps backward. This backward devolution was caused by planetary disasters, and this is shown by the forgotten worldwide pyramid culture. The globe is dotted with pyramidal structures in Russia, China, Kosovo, Sudan,

Germany, Italy, South America, Mexico, and more. The most reasonable explanation for such similar construction is that they were built when there was still a Pangean supercontinent. We do not have any true dating for when and how the supercontinent separated, and I offer no supporting evidence beyond a theory.

"If you knew the magnificence of the 3,6, and 9, then you would have a key to the Universe!"

Nikolai Tesla

www.ingramcontent.com/pod-product-compliance
Lightning Source LLC
Chambersburg PA
CBHW071459160426
43195CB00013B/2158